## WIRELESS-WISE FAMILIES

Lyn McLean is Australia's foremost consumer educator on the issue of electromagnetic radiation (EMR). She has been monitoring and writing on the subject for more than 20 years. Lyn is the author of *Watt's the Buzz*, *The Force*, and *Wireless-wise Kids*, and is the publisher of the quarterly periodical *EMR and Health*. She has represented the community on various national committees and is director of EMR Australia.

LYN McLEAN

# WIRELESS-WISE FAMILIES

## what every parent needs to know about wireless technologies

SCRIBE
Melbourne • London

Scribe Publications
18–20 Edward St, Brunswick, Victoria 3056, Australia
2 John Street, Clerkenwell, London, WC1N 2ES, United Kingdom

First published by Scribe 2017



Book design and illustrations by Scribe

Typeset in Adobe Garamond 11/17pt by the publishers

Printed and bound in China by Imago

Scribe Publications is committed to the sustainable use of natural resources and the use of paper products made responsibly from those resources.

9781925322248 (paperback)
9781925548600 (e-book)

A CiP record for this title is available from the National Library of Australia.

scribepublications.com.au
scribepublications.co.uk

For Jacqueline and Alexandra

# Contents

# Introduction

We love them, we depend on them, we fill our homes with them. We put them next to our sleeping babies and give them to our toddlers and children to play with. We use them for work, for entertainment, and for conducting our relationships.

Wireless devices are such a feature of modern life that chances are you use them, too.

But how much do you know about these devices and the radiation they emit?

Did you know, for example, that your use of wireless devices can affect your children, your unborn children, and even your chances of conceiving?

Did you know that wireless radiation has been shown to affect performance, health, mental health, relationships, and sleep?

Did you know that many world authorities recommend reducing exposure to this radiation?

If you're worried, you're not alone.

In this book, you'll learn what you need to know about wireless radiation, how you and your family can use wireless devices more safely, and why. You'll have a better understanding of your family's exposure and will discover tips for reducing exposure if you choose. You'll also find useful suggestions for understanding and managing the impacts of wireless technologies on your home, your life, and the lives of your family.

'Smart' families are exposed to wireless radiation from 'smart' devices.

'Wise' families make informed decisions about using wireless technologies safely.

CAFÉ
FREE WI-FI

Chapter 1

# What Is Wireless Radiation?

Wireless devices emit radiofrequency radiation (RFR), sometimes known as wireless radiation, which is the term used in this book.

This radiation is also sometimes called high-frequency radiation, to differentiate it from the lower frequencies emitted by electrical equipment such as power lines.

Wireless devices work by sending and receiving information that is superimposed onto a radiofrequency signal. You could imagine that the wireless radiation is a train travelling from point A to point B. The information (voice, text, or image) is like the passengers it carries. If you block the journey of the train, there is no transfer of passengers. In the same way, if you block the radiation, there is no transfer of data. In other words, wireless networks rely on radiofrequency radiation to carry information.

Wireless networks are constantly transferring information from one wireless point to another, for example:

- from a mobile phone (cell phone) to a base station (cell tower)

- from a cordless handset to its base
- from a laptop to a modem
- from a tablet to a router.

One of the features of wireless radiation is its frequency, which is measured in units called hertz, usually in the range of megahertz (MHz) or gigahertz (GHz). Different technologies operate at different frequencies. Mobile phones generally operate at frequencies from 700 MHz to 2,600 MHz (2.6 GHz). Microwave ovens operate at 2.45 GHz — the same frequency used for wi-fi.

The power of the radiation from a wireless device will depend on the area that the technology is designed to cover. A mobile phone might operate at low power if connecting to a mobile-phone base station nearby, but at much higher power if it were in a low-signal area like a lift or a rural area a long way from a base station.

How much exposure you receive from a wireless device will depend on two things — the power of its signal and the distance between you. The highest exposure occurs close to a wireless device, such as when your mobile phone is directly against your head or your body, or when you're standing close to a router or cordless phone. However, keep in mind that wireless radiation is nevertheless designed to transmit over distances — up to many kilometres in the case of some mobile-phone base stations.

Over the years, advertisers have coined a range of terms as alternatives to the word 'radiation', which has negative connotations. Conversely, they've invented negative terms to apply to radiation-free zones. Here are some of those terms and what they really mean:

- ‘energy’ — radiofrequency radiation
- ‘signal’ — radiofrequency radiation
- ‘reception’ — radiofrequency radiation
- ‘wireless’ — radiofrequency radiation
- ‘wi-fi’ — radiofrequency radiation
- ‘Bluetooth’ — radiofrequency radiation
- ‘smart’ meter/TV/appliance — one that emits radiofrequency radiation
- ‘coverage’ — an area covered in radiofrequency radiation
- ‘connect’ — send a signal of radiofrequency radiation to and from
- ‘black spot’ — a radiation-free zone

Would you be as pleased to take advantage of this offer if the sign said ‘Free radiation’?

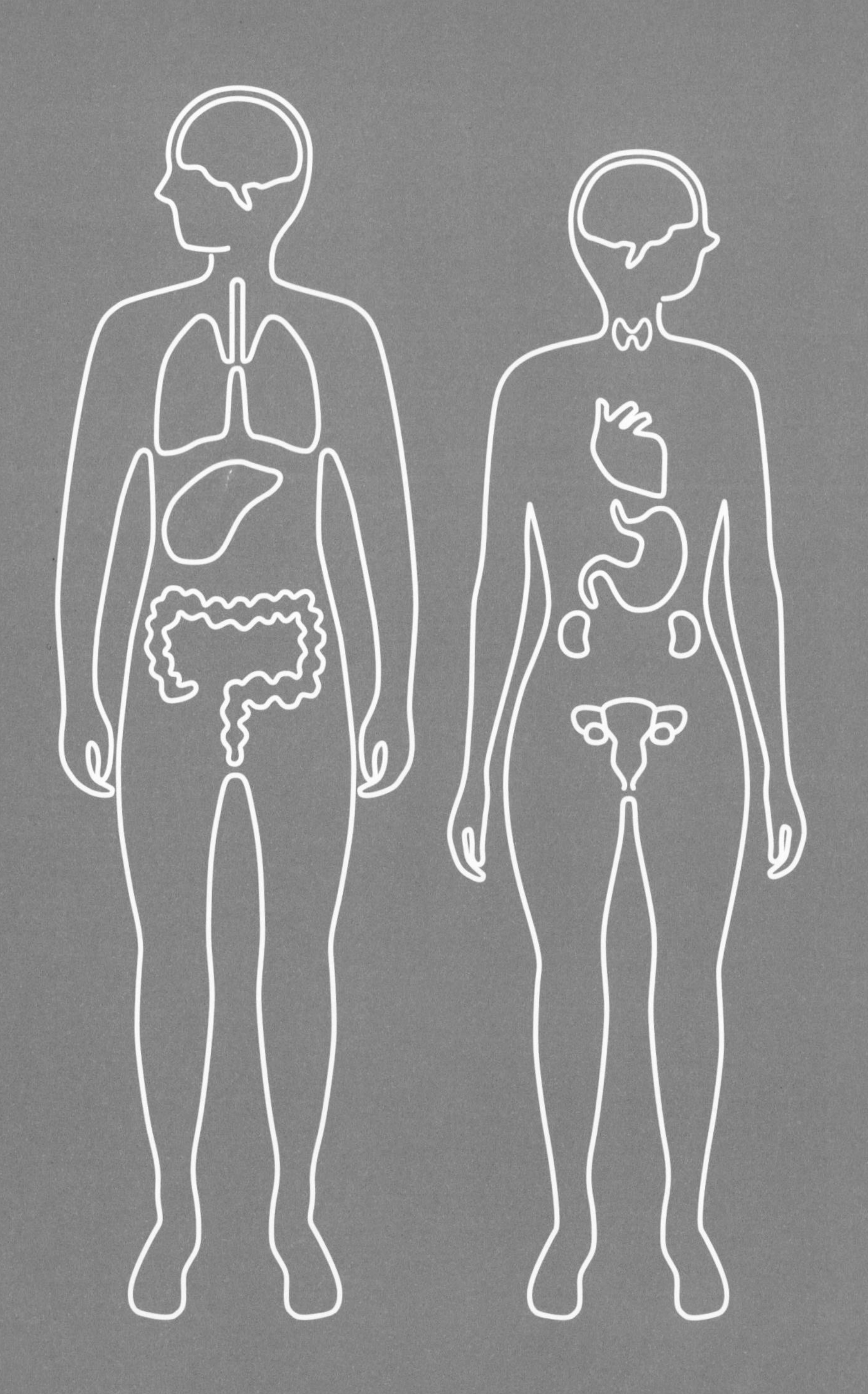

## Chapter 2

# Why Take Precautions?

When you bought the wireless devices in your home, you probably thought they were safe. They'd have to be safe to be sold on the market, wouldn't they? It's a reasonable assumption — but, unfortunately, it's not necessarily true.

Across the world, more and more concerns are emerging about the risks of wireless radiation, particularly for long-term exposure. Here are just some of the concerns raised so far.

## Cancer risks

In 2011, the International Agency for Research on Cancer (IARC) classified radiofrequency radiation as a Class 2B ('possible') carcinogen, in the same category as lead. IARC made this decision on the basis of two groups of studies (one by the Interphone team and one by Swedish oncologist Professor Lennart Hardell) both of which found higher risks of some brain tumours among heavy and long-term mobile-phone users, even though the phones they used complied with relevant standards.

Since then, several more studies have found evidence for brain-tumour risks from mobile-phone use.

One of these was the French study known as CERENAT. Its authors found that making one call a week for at least six months (which they called 'regular' mobile-phone use) didn't increase the tumour risk, but there was an increased risk of gliomas and meningiomas for people who had used a mobile phone for 896 hours altogether — equivalent to just 30 minutes' phone use a day over five years.

Professor Lennart Hardell and his team from Örebro University have found additional evidence that questions the safety of wireless radiation. They conducted a number of studies that found increased risk of gliomas, acoustic neuromas, and, to a lesser extent, meningiomas among people who had used mobile phones for more than ten years. In 2014, the team showed that long-term use of mobile phones decreased the survival rate for people with brain tumours.

In 2016, the US National Toxicology Program released the results of a study it considered too important to keep until the final results of its $25-million research program were published. It found that rats exposed to mobile-phone radiation had more malignant brain tumours, schwannomas, and DNA damage than unexposed animals.

Of course, not all studies have found increased brain-tumour risks — but some of these have been accused of serious design flaws.

A number of scientists believe that there is now enough evidence linking mobile-phone use with brain-tumour risks to categorise mobile-phone radiation as a Class 2A ('probable') carcinogen.

## Scientists have recognised risks

There are now thousands of peer-reviewed scientific studies in which scientists have found harmful effects from exposure to radiation levels that are well within international safety standards. These effects include:

- DNA damage
- chromosome damage
- changes to enzymes
- interruption of the cell cycle and cell communication
- reduced immunity
- sperm damage
- effects on hormones
- free-radical damage
- cellular stress
- changes to brainwave patterns.

For almost two decades, scientists and medical practitioners have expressed concerns about the risks of wireless technology in the papers they've published or in group statements and resolutions.

These concerns culminated in an Appeal to the World Health Organization (WHO) and the United Nations (UN), signed by 191 scientists from 39 nations on 11 May 2015. The text of the Appeal said: 'Effects include increased cancer risk, cellular stress, increase in harmful free radicals, genetic damages, structural and functional changes of the reproductive system, learning and memory deficits, neurological disorders, and negative impacts on general well-being in humans.' The Appeal called on the UN and WHO to take action to protect children and pregnant women, strengthen standards, encourage the development of safer technologies, and educate the public and medical professionals about the effects of wireless-radiation exposure. Since its launch, even more scientists have endorsed the Appeal.

## Courts have recognised risks

In several countries, courts have made determinations recognising that exposure to wireless radiation could be a health risk, even though those exposures complied with relevant standards.

### ITALY

In 2012, Italy's Supreme Court accepted that the brain tumour suffered by 60-year-old businessman Innocente Marcolini was caused by the radiation from his mobile phone. Marcolini had used his mobile phone for an average of five or six hours a day for about 12 years. He claimed that his phone use led to the development of a tumour on his trigeminal cranial nerve (responsible for sensations in the face), which was located close to the position of his phone — and the court agreed. Although the tumour was not cancerous, it was life-threatening and required removal by surgery, which left him with partial paralysis of the face, and ongoing pain.

In 2017, two Italian courts found a link between mobile-phone radiation and acoustic neuromas. A court in the city of Ivrea ordered the state's social-security agency to pay Roberto Romeo an annual pension of 6,000 to 7,000 euros for the injuries he sustained using a company phone for three hours a day for 15 years. At almost the same time, a court in Florence ordered the Italian Workers' Compensation Authority to compensate a worker who developed an acoustic neuroma after using his mobile phone for two to three hours a day for ten years.

### AUSTRALIA

In 2013, an Australian court recognised a link between the symptoms experienced by an Australian scientist and his exposure at work. The Administrative Appeals Tribunal of Australia ordered the federal government to pay compensation to Dr David McDonald for injuries he developed at work when exposed to electric and electronic equipment. Dr McDonald, a principal research scientist at the Commonwealth Scientific and Industrial Research Organisation (CSIRO), had advised his employer that he suffered from the condition of electromagnetic hypersensitivity (symptoms experienced when exposed to electromagnetic fields, including wireless radiation). When the CSIRO required him to undertake a trial of electric and electronic equipment, he developed severe symptoms within minutes — nausea and migraines that lasted for days.

### FRANCE

In 2015, a Toulouse court recognised that electromagnetic fields, including wireless radiation, were responsible for the symptoms of a 39-year-old woman who suffered from the condition of electromagnetic hypersensitivity. The court awarded a monthly disability payment of 800 euros to Marine Richard, who lived in a remote area, away from electromagnetic signals.

### ISRAEL

In 2014, the Tel Aviv-Yafo District Court accepted a settlement in which mobile-phone operators are required to provide consumers with information about safer mobile-phone use — including keeping phones at a distance from the body — and must provide an inexpensively priced hands-free kit with every mobile phone sold.

### SPAIN

The Spanish High Court awarded compensation to a university professor who suffered from electromagnetic hypersensitivity, among other problems. Meanwhile, a court in Castellón awarded a disability payment and carer's pension to a man who suffered from electromagnetic hypersensitivity.

### USA

On 12 August 2015, a 12-year-old boy, unnamed for privacy reasons, and his parents filed an injunction against the Fay School in the Massachusetts District Court. The boy developed symptoms of headaches, itchy skin, and rashes soon after an industrial-strength wi-fi system was introduced at the school. The parents alleged that the school refused to accommodate the boy's symptoms and disregarded his rights under the Americans with Disabilities Act.

How the court will decide in this case remains to be seen. However, already the local school district is taking action.

### INDIA

In 2017, the Supreme Court of India ordered that telecommunications company BSNL deactivate a mobile-phone base station because the radiation it emitted caused a plaintiff's cancer. Harish Chand Tiwari told the court that he'd developed Hodgkin's lymphoma as a result of 24-hour exposure from the tower for 14 years.

## Governments have recognised risks

Because of the many concerns about the safety of wireless radiation, many countries have introduced precautions to protect users. So have the governments of some regions and cities. These precautions include warnings for reducing exposure to mobile-phone radiation — especially for children — and introducing stricter standards. Some notable precautionary regulations have been introduced by the governments of France, Belgium, and Israel. You'll see more about these precautions later.

For more information
see page 101

## Authorities have recognised risks

A steadily increasing number of international authorities have suggested precautions for reducing exposure to wireless radiation. These include the Council of Europe, the Japan Federation of Bar Associations, peak paediatric bodies, medical associations, cancer institutes, associations of teachers, and various schools.

## Insurers have recognised risks

Some insurance companies have undertaken not to provide cover for injuries developed as a result of exposure to wireless radiation. For example, A&E Insurance has excluded cover for any problems caused by electromagnetic radiation.

Swiss Re recognised the risk of 'electrosmog' for insurers as early as 1997 and said, 'On the basis of present knowledge alone, it must be expected that plaintiffs will win suits dealing with this issue.' In 2013, it rated the unforeseen consequences of electromagnetic fields (including wireless radiation) as potentially having a high impact and said, 'If a direct link between EMF and human health problems were established, it would open doors for new claims and could ultimately lead to large losses under product liability covers. Liability rates would likely rise.'

Australian telecommunications giant Telstra admitted in its 2004 annual report that the company had been able to obtain only limited insurance cover against the risks of exposure and even that was diminishing. The company now self-insures.

## Mobile-phone companies have recognised risks

Mobile-phone companies are recognising potential risks of holding mobile phones directly against the body for calls and are including information in their user manuals for safer phone use (see page 21).

In Australia, Telstra has been texting its customers with a link to information on its website about how to reduce mobile-phone exposure. In one 12-month period, it reported sending 15 million of these messages.

The multinational telecommunications company Orange has published 'best practice' recommendations for reducing exposure to radiation on its website. These include recommendations to use a hands-free kit, to text rather than call, to keep phones away from the body of pregnant women, and to use phones in good reception areas.

## Problems with standards

The radiation from all wireless devices on the market must comply with relevant radiation standards. However, international guidelines and standards may not be adequately protecting the public. They have been designed to prevent a very limited number of short-term, acute heating effects of radiation, with an assumed safety factor added. As mentioned before, thousands of studies have found non-beneficial effects at levels below the ones they mandate.

Certainly some studies have failed to find harmful effects from radiation — and that's especially true of studies that have been conducted on animals and cells. However, there's evidence that these studies might not have been conducted properly in the first place. In 2015, Gernot Schmid and Niels Kuster, scientists from Austria and Switzerland respectively, published the results of a groundbreaking study in which they calculated that a mobile phone can expose tissues to over 40 Watts of radiation per kilogram (40 W/kg). However, when they looked at studies that had been published from 2002 onwards, they found that many of them had exposed cells to an average of approximately 2 W/kg. In other words, the results of these studies were not reliable for estimating risk.

So standards that have taken these studies into account are also highly questionable.

Anyone who tells you that wireless radiation is safe is telling you, not a fact, but an assumption. They are assuming that it should be safe if it complies with relevant standards, assuming the assumptions that underlie those standards are correct, and assuming the way products are being tested for compliance with those standards is correct.

That's a lot of assumptions.

The fact is that not only do we *not* know wireless devices are safe, but also there's evidence they're not.

It's simply not possible to state that wireless radiation is safe. Radiation-emitting devices are not tested for safety (only compliance) before being released on the market. All research on the effects of exposure is done afterwards — on the people using the technologies. It stands to reason that health effects from using wireless devices may not be immediately obvious. It takes around 40 years for brain tumours to show up, for example, so it may be decades before we know the full impact of exposure.

In short, no one knows what levels of long-term exposure is safe for any individual — and it may be decades before we do.

In the meantime, it's up to each of us to decide just how much exposure we are willing to accept as individuals and for our families. In the following pages, you'll find some tips about how to reduce your exposure if you choose — but ultimately, the decision about whether or not or how far to do so is up to you.

## NOTES

Chapter 3

# Mobile Phones

When someone in your family holds a mobile phone against their head, the level of radiation the phone emits needs to be strong enough to travel to the nearest base station, which could be anything from metres to kilometres away. The radiation penetrates the walls of your house and any solid objects that are in the way. Needless to say, the signal also penetrates the person's skull and is absorbed by their brain.

How much radiation their brain absorbs will depend on the person's size and age. Children's brains absorb much more radiation than those of adults, and so does their bone marrow.

Similarly, if someone in your family holds a phone next to some other part of their body — for example, in a bra, shirt pocket, or trouser pocket — then the radiation penetrates the closest part of their body — their breast, hip, or uterus or testes.

The important thing for you to know is that when your family's phones were tested for compliance with relevant radiation standards, they were found to comply as long as they were held at a distance from the head. They were never tested for compliance or found to comply if held directly against the head or the body.

You can check this is the case by looking at the User Guide for your mobile phone. The same information is carried on many of the devices themselves. For instance, on an iPhone, you can tap Settings > General > About > Legal > RF Exposure to read: 'To reduce exposure to RF energy, use a hands-free option, such as the built-in speakerphone, the supplied headphones or other similar accessories.' Until very recently, the message also advised users, 'Carry iPhone at least 5mm away from your body to ensure exposure levels remain at or below the as-tested levels.' This has been replaced with the simple assertion that testing was carried out 'with 5mm separation'.

This means that a person who holds a phone directly against their head or body, which most people do, is exposed to more radiation than allowed by international standards and guidelines.

Here's what happened to four young women who did just that.

A 21-year-old woman consulted a doctor for a bloody discharge from her left nipple and was diagnosed with breast cancer. She had carried her phone in the left side of her bra, right next to the position of the tumour, for several hours each day. Another 21-year-old developed tumours in the part of her breast adjacent to the position of her phone, which she'd been carrying for eight hours a day for six years. A 33-year-old developed tumours of her right breast immediately underneath the positon in which she'd stored her phone in her bra on and off for eight years. A 39-year-old developed tumours in her right breast where she'd placed a phone she used with a Bluetooth device while commuting to work over a ten-year period. None of these women had any family history of breast cancer or any other breast-cancer risks. All tumours were adjacent to the position where the phone was stored. All women underwent mastectomies.

As I mentioned earlier, there are concerns about how safe it is to use mobile phones for long periods of time. Scientists have found increased risks of brain tumours for long-term mobile-phone use (generally ten years or more) or heavy phone use. Some of those studies have reported no increased risks of 'regular' mobile-phone use — but they have defined this as using a mobile phone for just one call a week for six months. That's anything but regular use in today's society, where it's not uncommon for people to spend hours on their phone each day.

Because radiofrequency radiation has been classed as a 'possible' carcinogen — and some scientists think there's enough evidence for it to be classed as a 'probable' carcinogen — many authorities have recommended that people reduce their exposure to mobile-phone radiation.

Many of these authorities express particular concern about the use of mobile phones by children, who are generally thought to be far more vulnerable to it than adults.

**For more about children and mobile phones, see page 53.**

**For more about authorities' recommendations, see page 101.**

## Risk and precaution

If you'd like to reduce your exposure and your family's exposure to mobile-phone radiation, here are some useful suggestions — including some behaviours to avoid.

### LEAST WISE

- Spending long periods on calls with the mobile phone directly against your head.
- Letting children have unlimited use of mobile phones from an early age.
- Using a mobile phone when you drive — apart from exposing all the occupants of the car, it increases the risk of accidents.
- Using a mobile phone close to your baby, for example while breastfeeding ('brexting').

### WISE

- Don't hold your mobile phone directly against your head for calls.
- Don't carry your mobile phone next to your body while it is turned on unless it's in a shielded pouch.
- Limit the amount of time you spend on your mobile phone.
- Text rather than speak.
- Try to return calls on a landline.
- Don't use a mobile phone in a car, bus, or train.
- Don't use your phone in low-reception areas, including lifts.
- Use an air-tube headset for calls (i.e. don't use a headset with a wire, which can conduct the signal into your head, and don't use one with Bluetooth, which is just another type of wireless radiation).
- Use the speaker function.
- Don't use mobile phones for playing games, listening to music, or other unnecessary uses.
- Don't keep your mobile phone next to your bed at night when it's turned on.

- Don't use a mobile phone when pregnant or near a young child.
- Purchase mobile phones with as few functions as possible to discourage unnecessary use.
- Turn your mobile phone off when it isn't needed.
- Use a suitable mobile-phone shield to block mobile-phone radiation.
- Especially reduce children's exposure to mobile-phone radiation.
- Set a good example.

**WISEST**

- **Use a corded (not cordless) phone. Yes, it is still possible to buy a corded phone!**

## Mobile kids

There are good reasons to believe that children are more vulnerable to mobile-phone radiation than adults — as well as concerns about the effects of phone use on attention, behaviour, and learning, which you'll see more about later. So it makes sense to reduce their exposure as much as possible. Here are some ways you can help them do this:

- Don't give babies and children a mobile phone to play with.
- Avoid giving a child a mobile phone until it becomes absolutely necessary for them to have one.
- Teach children to keep their mobile phones turned off unless there is an emergency.
- Teach them responsible mobile-phone use.
- Don't allow children to have a mobile phone in their bedroom at night.

## Mobile-phone shields and protective devices

As people's concerns about mobile-phone radiation have risen, so have the number of shields and so-called protective devices on the market — some of which work and some of which don't.

Products that do work are those that cause a measurable reduction in exposure to the user. This is usually demonstrated in SAR (specific-absorption rate) tests conducted by reputable laboratories, so look for the test reports before buying. This category of product includes mobile-phone pockets and cases with screening on one side (to block radiation to the user's head) but not on the other (allowing the phone to connect with a base station). However, no product can block all radiation from a mobile phone — if it did, the phone would not be able to connect with the network.

There is also a huge range of products claiming to harmonise or neutralise mobile-phone radiation. Unfortunately, these products make no difference to the amount of radiation emitted by the phones in SAR tests, and their claims can't be scientifically tested.

## Strategies for reducing your family's exposure

If your family is like most, you may have become so used to using your mobile phones that family members are exposed to more radiation than you would like. If that's the case, here are some strategies that might be helpful in reducing exposure and cutting down on mobile-phone overuse.

### BEDROOMS

Make the bedrooms in your home places of rest and safety.

- Take the mobile phones out of all bedrooms at night.
- If you need an alarm clock, try a battery-operated one — they have no wireless emissions and no magnetic fields.

### PLAN AHEAD

Being organised avoids the need for many panicked calls for emergencies such as *do you want one kilo of sausages or two?*

- Write a shopping list.
- Work out where to meet in advance.
- Before they leave home, make sure you know where your children are going to be so you don't need to keep ringing them to find out!

### SET UP GOOD ALTERNATIVES

There are times you might need to use your mobile phone, but there are plenty of times you can do just as well without it.

- Take a book to read on the train.
- Retain or renew your landline so that you can be called without using the mobile network. Give people your landline number rather than your mobile number.
- Install a corded phone so that it's available when you need it — and use it for most of your phone calls.
- Set up an answering machine or voicemail on your phone line.

If you miss someone's call, you can call them back later.

- Set up a desktop or laptop with wired internet connection — and make sure the wireless networking is disabled — so that you can use this computer rather than your phone for accessing the internet.

### THINK ABOUT OTHER PEOPLE'S EXPOSURE

- Ring people on their landline rather than their mobile number.
- Avoid using your phone in crowded places, such as public transport, where you expose others.

### CONNECT MEANINGFULLY

Use the time you save on unnecessary phone activities to connect face-to-face with your family and friends! This is a good tip not only for avoiding wireless radiation, but also for living a more mindful, fulfilling life.

NOTES

Chapter 4

# Cordless Phones

The radiation from cordless phones is similar to that from mobile phones, so all that's been said about the latter applies to cordless phones as well.

In fact, cordless phones may be even more of a problem than mobile phones. That's because:

- Many people spend longer on their home or office phone than they do on their mobile phone, often thinking it's safer. It's not!
- Most cordless phones don't have adaptive power control, so they don't adjust their power for high- and low-reception areas. This means they are operating at high power all the time — and that means more radiation is absorbed by the users' heads when they make calls.
- Cordless phones emit radiation from both the handset and the base. In many cases, the base of a cordless phone transmits a signal 24/7.

Professor Lennart Hardell has found that long-term use of a cordless phone increases the risk of some types of brain tumours. For example, in a study published in 2015, he found double the risk of gliomas among the people who had used cordless phones for the longest — and risk was greater for people who had begun using such phones before the age of 20. In another study, he found that people who had used a cordless phone for more than 20 years had six-and-a-half times the rate of acoustic neuromas.

---

**SUGGESTION**

Try this experiment.

Make a call on your cordless phone. As you talk, walk out of your house and as far down the road as you can before the call cuts out.

That's how far the radiated signal from your cordless phone and cradle reaches.

If it reaches beyond your home in every direction, then it's certainly penetrating all the rooms of your home and exposing everyone inside to some wireless radiation.

---

## Don't be fooled

There's no such thing as a 'safe' cordless phone. Some retailers sell cordless phones that they claim are 'safe' on the basis that the cradle does not transmit a wireless signal 24 hours a day. However, this does not make the phone *safe*. You and your family would still be exposed to wireless radiation whenever you speak on the phone, just as you would if you use a mobile phone. Perhaps even more so.

## Increase your freedom

Perhaps you like having a cordless phone for the freedom it gives you to move around while you talk.

You can increase the freedom you have from a corded phone in several ways.

- You can buy a corded phone with a speaker function, so that you can move around the room as you talk.
- You can buy a longer cord for your corded phone — that way you can talk on the phone where you like.

Of course, a corded phone will give you the most important freedom of all — freedom from radiation.

## Risk and precaution

If you do use a cordless phone, here are some thoughts about how use affects exposure.

**LEAST WISE**

- Spending long periods using a cordless phone against your head.
- Allowing children to spend unlimited time on a cordless phone.
- Locating a cordless phone near a bed or other high-use area.

**WISE**

- Don't hold the cordless phone directly against your head for calls — use speaker function instead.

- Limit the amount of time you spend on your cordless phone.
- Don't locate a cordless phone next to a bed, on a desk, next to a lounge, or close to a kitchen work area.
- Don't allow children to use a cordless phone.

**WISEST**

- **Use a corded phone (with no wireless function). If you want to keep your hands free, buy a phone with a speaker function so you can work while you talk.**

## NOTES

Chapter 5

# Wi-fi

Wireless technologies have made such inroads into consumer devices that they're now finding their way into homes in droves.

As well as the well-known wireless phones, tablets, computers, printers, TVs, and baby monitors, there are now, or soon will be, wireless coffee makers, washing machines, clothes dryers, refrigerators, vacuum cleaners, and toys.

Wireless devices are being used so often that many people are addicted to them — more on that later. People are so used to sending texts and emails, that now they're even able to do it in their sleep! Sleep specialists have reported cases of people sending texts or emails while asleep — some of them coherent messages — with the people having no knowledge of doing so when they wake up.

**SUGGESTION**

Do you know how many wireless devices are present in your home?

Make a list of all the wireless devices in your home.

..........................................................................................

..........................................................................................

..........................................................................................

..........................................................................................

..........................................................................................

..........................................................................................

..........................................................................................

..........................................................................................

..........................................................................................

..........................................................................................

..........................................................................................

..........................................................................................

..........................................................................................

..........................................................................................

..........................................................................................

..........................................................................................

..........................................................................................

..........................................................................................

..........................................................................................

..........................................................................................

## Wi-fi and health

Of course, all these devices emit wireless radiation, some of them continuously.

Because wi-fi is a relatively recent technology, we don't know the full extent of its impacts, particularly on people whose exposure begins as children. However, studies on animals have shown damaging effects such as these:

- changes to microRNA — molecules that regulate the expression of genes — which could be linked with neurodegenerative diseases
- damage to the reproductive system
- interference with pregnancy or deformity of the embryo
- brain damage
- reduced growth and delayed puberty
- damage to the eye lens
- reduced liver function
- poorer learning and memory
- cardiovascular damage.

It's also likely that wi-fi exposure could interfere with fertility. Scientists from Argentina took samples of semen from 29 men and exposed half of it to wi-fi from a laptop for four hours. The exposed sperm had less motility than the unexposed sperm. In other words, they were less likely to navigate the route to the egg to achieve fertilisation. Since then, other studies have found similar results.

Of course, it's not possible to know the full effects of wi-fi radiation on people's health — because not enough time has yet passed to see the effects of long-term exposure. It will only be when the current generation of wi-fi users has been exposed to this radiation for many years that scientists will definitively understand the long-term effects.

However, the research on animals and sperm has already provided enough evidence to justify calls for precautions.

Professor Yuri Grigoriev, head of the Russian National Committee on Non-ionizing Radiation Protection, is a world authority on wireless radiation. He has described wi-fi radiation as 'an uncontrolled global experiment on the health of mankind'. 'We should stop telling [everyone] that Wi-Fi is harmless,' he wrote. 'We should better be honest and say that "we do not know what long-term effects might be."'

---

**WHAT ABOUT EYESIGHT?**

Children are spending so much time indoors, peering at screen-based devices, that it's affecting their eyesight. Levels of myopia (short-sightedness) are on the rise in many countries and one study found that over half the students in 60 primary and middle schools suffered from the condition.

---

## Useful things to know

- You can have access to the internet with wired connections (e.g. ethernet cables). Using a laptop with wired connections avoids wireless radiation — but make sure the you turn off the wireless on your devices!

- Higher levels of radiation are present when a wireless device is actually downloading or sending information, but radiation is present all the time that the device is wirelessly connected.
- No, you can't easily shield the radiation from wireless devices because, as shown earlier, it's the radiation that carries the information. If you successfully blocked all the radiation from a wi-fi device, such as a tablet, it would no longer connect to the internet. Similarly, if you successfully shielded a wi-fi router, it would no longer connect to your wi-fi devices.
- On some modems, you can turn of wi-fi when you are not using it — but be aware that some modems turn themselves back on automatically — so look for modems that don't.

---

**DE-STRESS**

Wireless radiation triggers a stress reaction in the body. But did you know that an effective antidote to stress is spending time in nature? Scottish scientists found that people who were walking in the natural environment had lower levels of frustration and arousal, and described their walks as 'mood-enhancing'.

---

## Pros and cons

The benefits of wireless technology is that there are less cords — less to trip over, less to pull out, and less to yank your expensive device from its perch. The cost of the technology is that it exposes the family to radiofrequency radiation.

Which would you rather have, the cord or the radiation?

---

**JUST HOW SMART IS YOUR SMART TV?**

Did you know that smart TVs don't just display television shows or internet content? They can also pick up information about homeowners. In early 2017, New York resident Joshua Siegel took legal action against Samsung, claiming that its smart TVs record conversations that take place in front of the TV and may share that information with a third party.

---

## Risk and precaution

If you'd like to reduce your family's exposure to wi-fi, here are some ideas to consider.

**LEAST WISE**

- Spending long periods using wi-fi.
- Locating the wi-fi modem or router in a high-use area such as a bedroom, on a desk, in the kitchen, or near the lounge.
- Allowing children to have unlimited use of wi-fi.
- Buying as many 'smart' appliances as you can.
- Buying or creating a smart house.

### WISE

- Turn off the wi-fi when it's not in use. Because some modems automatically turn the wi-fi back on, you might consider buying one that allows wi-fi to be turned off permanently.
- Locate the modem or router away from high-use areas.
- Limit children's use of wi-fi (e.g. by encouraging them to do their homework on a computer with wired internet connections).
- Buy devices that can be plugged into a wired modem (e.g. a laptop) rather than those that can only be used wirelessly (e.g. a tablet).
- Don't use devices that have wi-fi — such as laptops and tablets — on your lap.
- Use your tablet in airplane mode as much as possible.
- Keep away from your wireless device when you're downloading a game or program from the internet, and play or watch it later with wi-fi turned off.
- Keep wireless devices out of bedrooms.
- Don't use wi-fi devices in low-reception areas or in cars, buses, or trains.
- Before Christmas and birthdays, let relatives know you'd prefer your child to receive non–wireless-radiation-emitting presents.

### WISEST

- **Use a wired computer with wired internet and turn off the wireless function.**

## THINGS I'D LIKE TO DO WITH MY SCREEN-FREE TIME

## NOTES

Chapter 6

# Digital Dads and Mobile Mums

If you're a digital dad or a mobile mum, the safety of your child or children is probably utmost in your mind.

So have you considered the effects of wireless radiation?

Did you know, for example, that your own use of wireless devices can affect your children and your unborn children? What's more, it can affect your chances of becoming a father or a mother in the first place.

If you're planning a family, here are a few things you need to know.

## Digital dads

Men, if there's one thing you need in order to father a child, it's healthy, active sperm. That's because a sperm's journey from launch to touchdown is so challenging that only one in around 300 million is likely to succeed. Anything that tips the odds against them is likely to contribute to infertility — which has been rising in many countries around the globe and currently affects about 14 per cent of couples.

For years, scientists have been finding that mobile-phone radiation has harmful effects on sperm. In 2014, scientists analysed the results of ten previous studies and found that men's exposure to mobile-phone radiation was indeed linked with reduced sperm motility and viability, and some studies found effects on sperm concentration.

From various experiments, there's also evidence that wi-fi radiation could be tipping the odds against fertility:

- Rats exposed to wi-fi radiation showed negative effects on sperm and testicular structure. The longer the exposure, the worse the effects.
- Rats exposed to wi-fi had lower sperm counts.
- Cell death occurred in the testes of exposed rats.
- DNA damage was found in the testes of growing rats.
- Exposed rats developed defects in the sperm head and had smaller testicular characteristics than normal.
- Exposed mice had lower sperm counts, less sperm viability, and degeneration of their seminiferous tubules (where sperm develop).
- Men with erectile dysfunction carried their mobile phones in their pockets, turned ON, for longer periods of time than men without erectile problems.

## Tips for dads

- Don't carry a mobile phone in your hip pocket or belt while it's turned on.
- Don't use a laptop on your lap, especially when it's in wireless mode.
- Reduce your exposure to mobile and wi-fi radiation (see pages 21 and 37).
- Don't ring the mother of your child on her mobile phone!

## Mobile mums

For mums and intending mums, the message is similar — mobile phones and wireless devices can affect the unborn child.

When pregnant rats were exposed to mobile-phone radiation, their offspring were shown to have developmental problems of one kind or another. These included abnormal kidney development, changes in the development of the cerebellum (hindbrain), loss of cells from the hippocampus (midbrain), structural damage of the cochlea (inner ear), testicular damage, liver damage, heart damage, and possibly damage to the thymus and spleen (both part of the immune system).

Scientists have also shown that exposing animals to wi-fi or mobile-phone radiation had harmful effects on their offspring.

- Exposure to wi-fi radiation suppressed pregnancy and could lead to deformity of the embryo if it survived.
- Exposure reduced hormones of pregnant rats and increased harmful oxidation in the uterus.
- Rats exposed to wi-fi radiation during gestation were slower to gain weight and had delayed puberty.
- Young rats exposed to wi-fi or mobile phone radiation before and after birth had kidney and testicular damage, and changes to the onset of puberty.
- Rats exposed to mobile-phone radiation in utero showed changes to Purkinje neurons, which are some of the largest and most complex neurons in the nervous system.
- Rats exposed to mobile-phone radiation in utero had poorer cognitive performance.

- Embryos of Japanese quails exposed to mobile-phone radiation had increased oxidative damage, including damage to DNA.

But what about the effects on human babies?

To help answer this question, two teams of scientists looked at a group of over 52,000 children in Denmark whose mothers were pregnant between 1996 and 2002 and whose children were aged seven at the time of the study. Some of these children had been exposed to mobile-phone radiation before birth, some after birth, some before and after birth, and some not at all. Children who were exposed both before and after birth had the highest rates of headaches and migraines, but those who had been exposed either before or after birth had a higher incidence of headaches than unexposed children.

Headaches are not the only problem linked to a mother's use of her mobile phone. A study of over 13,000 children found that children exposed to mobile-phone radiation before birth — and, to a lesser extent, after birth — had more behavioural problems by the age of seven than unexposed children, including hyperactive, ADHD-like behaviour. The authors also found behavioural problems in an even larger group of children some years later.

Similar behavioural problems were found a few years later in a large, multinational study of nearly 84,000 mothers and children from Denmark, South Korea, the Netherlands, Norway, and Spain. The researchers found that mothers who didn't use mobile phones during pregnancy — over a third of the sample — had children with the least behavioural and emotional problems. The more mobile-phone use mothers reported, the more likely their children were to develop hyperactivity, inattention, behavioural problems, or emotional problems.

To see whether a mother's mobile-phone use could, in fact, cause ADHD in her children, Dr Hugh Taylor, a professor in reproductive medicine at Yale University, conducted an experiment in which he exposed female mice to mobile-phone radiation throughout pregnancy. He found the offspring had more signs of hyperactivity and less memory capacity. 'We have shown that behavioral problems in mice that resemble ADHD are caused by cell phone exposure in the womb,' Dr Taylor said. 'The rise in behavioral disorders in human children may be in part due to fetal cellular telephone irradiation exposure.'

In a completely different type of study, scientists in Iran looked at mobile-phone habits of mothers of children with and without speech problems. They found that children with speech problems tended to have mothers with higher mobile-phone use.

## Tips for mums

If you're planning a family or have a baby on the way, there are some things you can do to reduce the potentially harmful effects of wireless radiation.

- Don't carry a mobile phone close to the foetus.
- Don't rest a wireless laptop/tablet or electronic reader on your lap, especially when it's in wireless mode.
- Reduce your use of mobile phones (see page 21), cordless phones (see page 31), and other wireless devices (see page 37).

Chapter 7

# Wireless Kids

Today's child is exposed to more wireless radiation than ever. In fact, selling wireless devices for children has become a lucrative industry, and there are now wireless devices of almost every imaginable description. They include:

- wireless baby monitors
- mobile-phone apps for babies
- mobile phones and apps for kids
- tablets for kids
- iPotties
- interactive dolls
- GPS tracker devices
- even smart pyjamas!

How much time children spend on their mobile phones alone is hard to say — it seems to be growing every year. Many babies are now using wireless devices for up to an hour a day before they can walk or talk. Teenagers aged 15 to 19 spend at least three hours each day on social media. By the age of 16, teenagers can be using a mobile phone for approximately six-and-a-half hours a day. And approximately a quarter of teenagers are online 'almost constantly'.

Because the current generation is the first to be exposed to so much wireless radiation from so many sources and for such long periods of time, we simply don't know what the long-term effects will be on their health or wellbeing. However, we can assume that children are much more at risk from wireless radiation than adults, for many reasons:

- they have thinner skulls and so they absorb more radiation than adults
- their bone marrow absorbs about ten times more radiation than an adult's bone marrow
- as they grow, their cells are dividing, and this phase (mitosis) is particularly vulnerable to radiation
- their smaller heads resonate with the frequencies of wireless radiation
- their brains are still growing and maturing
- they have a potential lifetime of exposure
- they do not have the knowledge or skills to make informed judgements about using wireless technology.

Because of this vulnerability, many authorities in many parts of the globe have recommended precautions to reduce children's exposure to radiation.

And for good reason.

Although we don't know the health effects of long-term exposure to wireless radiation, there are already some worrying signs.

## Hyperactivity

We've already seen that young animals exposed to mobile phone and wireless radiation before birth had hyperactive and ADHD-like behaviours. We've also seen that children exposed to their mother's mobile-phone radiation before and after birth developed ADHD-like behavioural problems.

So it's no surprise that ADHD-like symptoms have been found in children who use mobile phones. Researchers studied a group of nearly 2,500 children in South Korea. They found that those who made more mobile-phone calls and were also exposed to lead had more chance of developing ADHD symptoms. They also found that children who spent more time playing games on their mobiles had higher rates of ADHD symptoms, irrespective of whether or not they were exposed to lead.

## Headaches

One of the symptoms that's often reported by adult mobile-phone users is headaches. And children who use mobile and cordless phones are reporting them now as well.

## Sleep problems and fatigue

Adequate good-quality sleep is important for health and performance, yet children's use of wireless technology is also often linked to sleep problems and fatigue. This is because children often stay up late at night using wireless devices and are often woken during the night by incoming messages. Some even sleep with their phones under their pillows.

## Brain tumours

The big concern is whether children who use mobile phones might be more at risk of the brain tumours that have been found among adults who are heavy and long-term mobile-phone users.

Australian brain surgeon Dr Charlie Teo believes this is the case. He told *60 Minutes* that he was worried about the number of children he was seeing with malignant brain tumours as early as 2009.

In his studies on brain tumours and mobile-phone use, Professor Lennart Hardell has also found higher risks among young people who began using mobile and cordless phones before the age of 20.

## Autism

It's also been suggested there could be a link between electromagnetic radiation and autism.

## Mental health

Children who spend long periods of time using wireless technology may have increased risks of psychological problems, including depression, as you'll see in the following chapter.

## Other problems

A number of other problems have been linked to children's use of screen-based technologies, whether or not they emit wireless radiation. However, because children's screen activities often involve wireless devices these days (whether a tablet or a smart TV), it's relevant to add such problems to our discussion. They include:

- vision problems
- appetite problems and obesity
- reduced levels of the hormone melatonin, leading to sleep and other problems
- cardiovascular disease and blood-pressure problems
- diabetes
- reduced fitness
- attention-deficit behaviours.

Apart from the health effects of children's use of wireless technologies, there are some worrying social problems. Online bullying is a major problem that's been widely reported. Suffice to say that wireless devices add a whole new dimension to the age-old problem of bullying, making it easier than ever. Now bullies can send unpleasant messages at the mere push of a button without the need to even be in the physical presence of their victims.

On top of all these potential risks, there's the problem of addiction. Today's children have become dependent on their wireless devices to the extent that they demonstrate withdrawal symptoms without them. Children as young as four have become so addicted that their parents are enrolling them in technology-addiction programs.

Despite all these problems, wireless devices play an important role in kids' education … don't they? More on that in the next chapter.

## Something to think about

The German telecommunications regulator, the Bundesnetzagentur, has banned the sale of the Cayla doll, a smart doll that can answer children's questions and play games with them using smart devices. The agency warns that smart toys with cameras or microphones that use insecure Bluetooth connections can transmit pictures and voice messages to and from children, potentially violating their privacy. Similarly, they could conceivably transmit sales pitches and other messages directly to children.

---

**SUGGESTION**

How many wireless devices do your children have?

Why don't you keep a record of how much time your child spends on wireless devices, with the wireless turned on, one day this week?

---

## WIRELESS DEVICES MY CHILDREN USE

| MY CHILD'S WIRELESS USE | DATE |
| --- | --- |
| **WIRELESS DEVICE** | **AMOUNT OF TIME** |
| | |
| | |
| | |
| | |
| | |
| | |
| | |
| | |
| | |
| | |
| | |
| | |
| | |
| | |
| | |
| | |
| | |
| | |
| | |
| | |
| | |

## Risk and precaution

**LEAST WISE**

- Locating radiation-emitting devices — such a baby monitor or mobile phone — close to a sleeping baby.
- Allowing the baby to play with your mobile phone or tablet as often as he or she wants to.

**WISE**

- Keep all wireless devices out of children's bedrooms at night.
- Don't allow children to use a cordless phone.
- Don't give a baby or toddler a mobile phone as a pacifier or toy.
- Don't put a wireless baby monitor in the baby's room.
- Don't give him or her a toy mobile phone or other wireless device — this sets up the expectation that it's OK to use these devices, and helps develop addictive behaviours.
- Don't use a mobile or cordless phone when you're breastfeeding or nursing your baby.
- Don't buy unnecessary radiating devices that are aimed at babies and children.
- Likewise, keep cots and beds away from wireless devices.
- Make sure there's a corded phone in the home so children don't need to use a mobile or cordless phone.
- Make sure children don't spend unnecessary time on wireless devices — for example, playing games.
- Use wired (not wireless) computers at home, including for homework.

- Turn the tablet onto airplane mode when your child is using it, and don't let your child use the tablet while it is downloading games or programs.
- If children must use a mobile phone, buy one that has the least functions and teach them safe use (see page 24).

**WISEST**

- **Keep wireless devices out of children's bedrooms.**
- **Make sure there are no wireless devices situated near children's bedrooms.**
- **Ideally, make sure there are no wireless devices in the home.**

## Other tips

- Avoid the risks of internet addiction by limiting children's use of wireless internet devices.
- Balance screen time with time spent in nature.
- Encourage the reading of books:
  - read to your child
  - provide an abundance of rich and varied reading material (libraries are a great source)
  - set aside times for reading (such as just before sleep)
  - read and discuss books with your child.
- Provide opportunities for your child to engage with other people socially, to develop interpersonal skills.

NOTES

Chapter 8

# Wireless Learning

Despite the fact that we simply don't know the long-term risks of wireless technology for kids, many parents and educators have rushed full-pelt into filling homes and classrooms with wireless devices on the assumption that it will benefit children's learning.

## Wireless classrooms

Let's look at a hypothetical situation in a typical classroom where students are engaged in internet activities using wi-fi-enabled devices. A boy in that classroom will be exposed to wireless radiation from the wi-fi router, from the wi-fi device that he's using, from all the wi-fi devices the other children in the classroom are using, from the smart phone in his pocket — and so on. So how much wireless radiation will that be?

That will depend on the strength of the signal from the router, how close the router is to the child, how much data is being downloaded, and how many children sitting close to him have their phones turned on at the time.

The levels of wireless radiation in that classroom will certainly comply with national and international standards — but they may be above levels at which scientists have identified harmful biological effects.

There are many situations in which children are developing unpleasant symptoms in wireless-enabled classrooms — symptoms that disappear when the children are away from wi-fi devices.

I've mentioned the case of the schoolboy in the United States who developed headaches, itchy skin, and rashes soon after an industrial-strength wi-fi system was introduced at his school. His parents took the Massachusetts school to court. As a result, the Ashland school district in Oregon introduced 'best practice' guidelines for reducing students' exposure to wireless radiation, to be displayed in every classroom.

In another case, a young British girl committed suicide in 2015 as a result, her mother said, of her allergies to wi-fi at school.

In Australia, some parents have chosen to homeschool their children after they developed symptoms when wi-fi was introduced in classrooms.

## Do wireless devices benefit education?

Wireless devices and wireless networks are often marketed as being wonderful tools for education. But how much benefit does this technology actually bring?

Every teacher knows that students can benefit from having internet access in the classroom to support their work. Most also know that students use wireless devices in the classroom to message each other, play games, distract themselves from work, and access inappropriate content.

There's a common assumption that giving children digital devices encourages them to spend time reading. But that's not the case, according to researchers from Western Australia. They found that children who were frequent readers preferred to read books rather than use electronic devices. Moreover, giving children mobile phones and multiple digital devices tended to reduce the amount of time they spent reading.

In Britain, some schools banned students from using mobile phones, and observed improvements in their results. In fact, the ban was found to improve academic performance, especially among low-achievers, and was calculated to add the equivalent of five extra days per year to their schooling.

Children's use of computers, wireless or otherwise, has been found to reduce academic performance. Playing computer games is, not surprisingly, linked with poorer literacy, and children given a home computer had poorer scores in reading and writing as well as more problems at school. The internet teaches children to read by skimming large quantities of information for superficial meaning, and this may come at the cost of the ability to access deeper levels of understanding. A report released by the OECD in 2015 found that moderate computer use at school slightly improved student performance, but that frequent computer use produced much worse results.

In his book *Glow Kids: how screen addiction is hijacking our kids — and how to break the trance*, Dr Nicholas Kardaras says that the use of wireless devices in education can interfere with the development of children's motor skills, logical thinking, literacy, and the ability to pay attention. Heavy users have been found to suffer from ADHD, screen addiction, aggression, depression, anxiety, and psychosis, he says. Worse, heavy use of screen-based devices can cause neurological damage similar to that caused by cocaine addiction.

Professor Yuri Grigoriev, head of Russia's peak radiation authority, found that children who use mobile phones may have more health and psychological problems than children who don't. He observed the progress of 147 children aged five to 12 years and found that children who used mobile phones had more fatigue; were less able to work at school and home; had poorer attention, accuracy, memory, and motor reactions; and had difficulty distinguishing sounds in speech.

Heavy use of wireless devices can not only lower academic performance, but also decrease happiness. In a group of undergraduates at the University of Kent in the United Kingdom, the heaviest users of mobile phones were less happy and performed less well academically than those who used their phones less. Similarly, Japanese students who were heavy mobile-phone users were more likely to suffer from 'depressed mood'. A group of young adults in Sweden who were heavy mobile-phone and computer users also had higher rates of depression, as well as higher stress levels and more sleep problems than normal.

Wireless radiation can also affect the brain's structure and function.

## Brain structure

Scientists in China were interested in how the brains of teenagers with Internet Addiction Disorder (IAD) compared to the brains of those who are not addicted. They recruited 18 volunteers with internet addiction, who used the internet from eight to 12 hours a day for approximately six days a week. They conducted MRI scans on the both groups of teens.

They found changes in both the grey matter (the part of the brain made up of closely packed neurons) and white matter (the deeper parts of the brain, made up of long nerve fibres that convey sensory information). The volunteers who were addicted to the internet had less grey matter in several areas of the brain responsible for cognitive control, selecting appropriate behaviour, goal-related behaviour, and higher cognitive functions (memory, behaviour, personality). The scientists found that the amount of grey matter in the brain decreased as the amount of time on the internet increased.

In addition, the scientists found changes in white matter in several parts of the brain responsible for memory, sensory information, and regulating emotions.

'Our results suggested that long-term internet addiction would result in brain structural alterations, which probably contributed to chronic dysfunction in subjects with IAD,' they wrote.

The Chinese researchers are not the only ones to cast doubt on the effects of heavy media use on the brain. Professor David Levy of the University of Washington refers to the 'popcorn brain' — a brain that is so acclimatised to the constant stimulation of online activity that it is unable to function properly at the slower pace of everyday reality.

## Safer school internet

If schools want students to access the internet for educational activities, there are safer ways of doing so than using wireless networks.

Using hardwired internet connections allows students to have the full benefit of internet access without the risks to health, behaviour, and learning that have been associated with wireless technology.

Many authorities have called for schools or preschools to avoid using wi-fi and to use wired rather than wireless internet, including the following:

- American Academy of Environmental Medicine
- Athens Medical Association
- Austrian Medical Association
- Cyprian government
- French government
- German government
- government of Navarra (autonomous region of Spain)
- government of South Tyrol (autonomous region of Italy)
- Israeli Department of Education
- municipality of Ghent, Belgium
- Ontario English Catholic Teachers' Association (Canada)
- Russian National Committee on Non-ionizing Radiation Protection
- Voice, the British trade union for education professionals.

---

**SAFE HOME INTERNET**

If you're a parent of a schoolchild, you can minimise your child's exposure to wi-fi radiation by making sure that homework is done on a computer with wired internet access (and wireless turned off) instead of a wireless device.

If your child needs a wireless device for school, a laptop is better than a tablet, because it can be connected to a wired modem for homework (with its wireless turned off).

---

## NOTES

## Chapter 9

# Wi-tech Relationships

How attached are you to your wireless devices?

Could you, or the members of your family, go for a day without using them?

If you answered *no*, you're not alone. Many people are so attached to their devices that they feel lonely, isolated, confused, or depressed without them. People are spending longer than ever on devices — staring at screens for eight hours a day, connecting to the internet before they get out of bed.

So, does people's relationship with their wireless devices change the nature of their relationship with others?

Absolutely — and here's the proof. Take a look in any restaurant where two people are sitting together and notice in how many instances one of them is using a wireless device.

## Wireless couples

That's because people are so addicted to their wireless devices, they can't go for long without checking them. A survey of American smart-phone owners found that almost 60 per cent said they could go for no more than an hour without checking their phones. They admitted checking their phones during meals, while driving, while on the toilet, and when they wake up during the night.

Needless to say, this trend is affecting relationships between couples. In fact, a 2010 survey of British couples found that one in ten spent more time talking to their partner by phone or email than they did in person. Even when they were together, over 40 per cent of couples communicated with their partner electronically rather than by speaking.

When people do get together to talk, the quality of their conversation is affected by the mere presence of a mobile phone. Scientists at the University of Essex asked pairs of strangers to chat for a ten-minute period, with and without mobile phones in the room. They found that simply having the mobile phone present affected closeness, connection, and conversation quality, and this was most apparent when the couples were discussing meaningful topics.

If wireless technology is changing our relationship with individuals, what is it doing to family relationships?

## Wireless families

The more time people spend engaging with their devices, the less time they spend engaging with each other.

Dr Jenny Radesky and her team set up an experiment where they observed 55 groups of people — containing at least one adult and one child under the age of ten — eating in fast-food restaurants. She found that 16 of the 55 adults were more absorbed in their phones than their children. Some were almost continuously involved in using their phone during the meal — either typing or swiping the screen. Those who used their phones to make calls kept some eye contact with their children, even if they appeared absorbed by their conversation.

Similarly, children are often too engrossed in their wireless devices even to notice when their parents — especially their fathers — come home from work.

In fact, family members are spending more time alone and less time in the same room as each other.

Melody Terras and Judith Ramsay decided to investigate what role the home environment played in children's online behaviour. They found that parents often complained about just how much time their children spent using the internet, but many of them were excessive mobile-phone users at home themselves! 'It is essential to raise parental awareness of the power of their own smartphone behaviors as it is often a case of "Do as I say rather than as I do,"' they wrote.

## Social skills

Not surprisingly, the skills needed for healthy social relationships seem to be at risk.

One of the most important ways that people communicate — more important even than spoken language — is body language. Yet today's children, spending less and less time with other people from an early age, have less opportunity to learn how to read the nuances of other people's behaviour.

One of the important qualities of successful relationships is the ability to understand another person's feelings and points of view — or empathy. It seems that the neurons involved in empathy are also involved in copying other people's behaviour. In other words, empathy appears to be learned by example. It's been suggested that empathy is in decline. Scientists found that empathy levels declined in college students after the turn of the century and are 40 per cent lower than levels in students 20 to 30 years earlier.

---

**SOMETHING TO THINK ABOUT**

Your child is learning how to relate to other people by mirroring your behaviour. What wireless behaviours do you think he or she is learning?

What behaviours about relating to others do you think he or she is learning?

Do you see any examples of your child mirroring your behaviour?

---

**SUGGESTION**

Why not keep a diary of your family's activities on a typical day. This could be a group project in which each member of the family records where they were and what they were doing. (To encourage honesty, everyone should write down the device they were using, rather than what they were doing on it!)

A chart for a family with school-age children might look something like this (see next page).

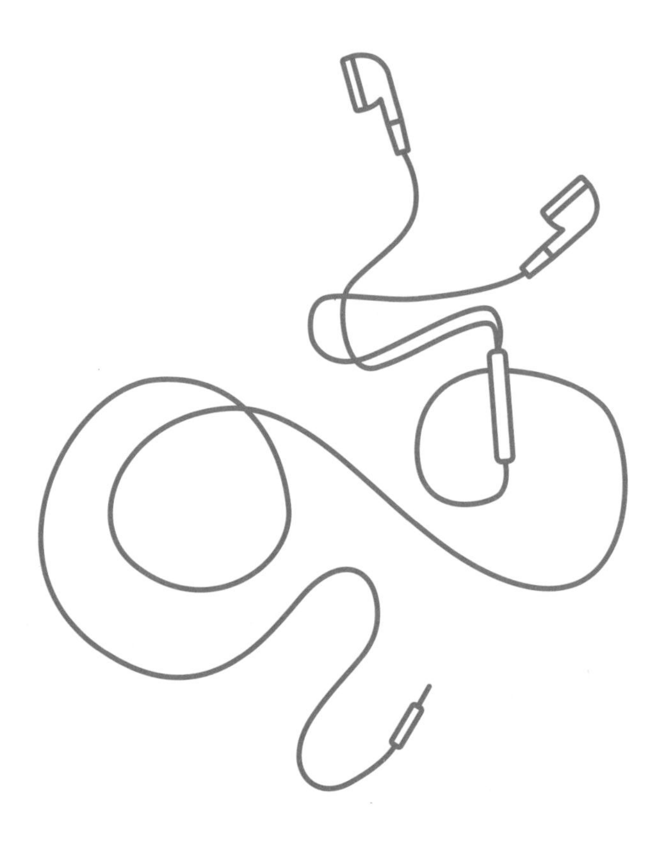

## Our family diary

| | PARENT 1 | | PARENT 2 | |
|---|---|---|---|---|
| TIME | WHERE | WHAT | WHERE | WHAT |
| 7.00 am | | | | |
| 7.30 am | | | | |
| 8.00 am | | | | |
| 8.30 am | | | | |
| 9.00 am | | | | |
| Recess | | | | |
| Lunch | | | | |
| 3.30 pm | | | | |
| 4.00 pm | | | | |
| 4.30 pm | | | | |
| 5.00 pm | | | | |
| 5.30 pm | | | | |
| 6.00 pm | | | | |
| 6.30 pm | | | | |
| 7.00 pm | | | | |
| 7.30 pm | | | | |
| 8.00 pm | | | | |
| 8.30 pm | | | | |
| 9.00 pm | | | | |
| 9.30 pm | | | | |
| 10.00 pm | | | | |
| 10.30 pm | | | | |
| 11.00 pm | | | | |
| Total time spent on wireless devices | | | | |
| Total time spent alone | | | | |

| CHILD 1 | | CHILD 2 | | CHILD 3 | |
|---|---|---|---|---|---|
| WHERE | WHAT | WHERE | WHAT | WHERE | WHAT |
| | | | | | |

**SOMETHING TO TRY**

You might like to try this internet addiction test?

1. Do you feel preoccupied with the internet (think about previous online activity or anticipate next online session)?
2. Do you feel the need to use the internet for increasing amounts of time in order to achieve satisfaction?
3. Have you repeatedly made unsuccessful efforts to control, cut back, or stop your internet use?
4. Do you feel restless, moody, depressed, or irritable when attempting to cut down or stop your internet use?
5. Do you stay online longer than originally intended?
6. Have you jeopardised or risked the loss of a significant relationship, job, or educational or career opportunity because of the internet?
7. Have you lied to family members, your therapist, or others to conceal the extent of your involvement with the internet?
8. Do you use the internet as a way of escaping from problems or of relieving a dysphoric mood (e.g. feelings of helplessness, guilt, anxiety, depression)?

---

Answering *yes* to questions 1 to 5 plus at least one other question may indicate the condition of Internet Addiction Disorder (IAD).

Of course, wireless technology doesn't have to interfere with interpersonal relationships, and can help build meaningful connections — but then, so can wired communications devices.

Here are some tips for doing this.

## Useful tips

### BE AWARE

Take a look at your relationship with your wireless devices. Notice how much time you spend on your devices each day and what you are using them for. Because most habits are unconscious behaviours, becoming conscious of your wireless activities is the first step in taking control of them.

Encourage other people in your family to do the same.

The chart on page 78 may be a useful starting point.

### ESTABLISH BOUNDARIES

Decide areas or times of your life that are wireless-free zones. They might include mealtime, bedtimes, and conversation times.

If you can, keep the use of wireless devices to a minimum and keep internet activities limited.

### PRIORITISE

Value the person you're with.

Every time you answer a mobile-phone call when you're with someone else, you're giving them the message: *this person who isn't even here is more important than you are.*

### FIND ALTERNATIVES

What better way to replace one habit than with another?

- Find enjoyable alternatives to spending time with your wireless device. Take up a hobby; take the dog for a walk; join a club; get back into reading.

- Encourage your children to engage with other people. Invite friends over after school; enrol your child in a sport.
- Find activities your family can engage in as a unit. Adopt a rescue pet; play board games; go on a family outing; visit friends; build a garden and care for it together.

---

**MY WIRELESS-FREE DECISIONS**

Make a list of the wireless-free zones, times and activities that you think are worth establishing here.

**WIRELESS-FREE ZONES**

..................................................................................

..................................................................................

..................................................................................

**WIRELESS-FREE TIMES**

..................................................................................

..................................................................................

..................................................................................

**WIRELESS-FREE ACTIVITIES**

..................................................................................

..................................................................................

..................................................................................

# NOTES

## Chapter 10

# Buying for Your Home

If you're buying new furnishings or appliances for your home, you have the perfect opportunity to choose ones that will help create the level of wellbeing for your family that you desire.

Here are some things to think about before you buy.

### Is it wireless or not?

You might think that most people would know whether the products they were purchasing were wireless or not, but that's not the case. So many of today's products unexpectedly have wireless function, it's easy to introduce wireless products to your home unintentionally. I've visited homes where people have been surprised to learn that their phone or their printer, which they'd connected with wires, was nevertheless emitting wireless radiation.

The words 'wi-fi', 'Bluetooth', 'smart', 'MHz', or 'GHz' on the packaging are indicators that a product uses wireless radiation.

## Check your options

Just because the electronics store has shelves of cordless phones, it doesn't mean there are not perfectly good corded options available. You may be surprised at the options available.

If you do decide to buy a wireless router, choose one that can be turned off when it's not being used and not one that will automatically turn itself back on.

## Choose connections

If you use a wired modem, choose devices that will connect with it easily. For example, a laptop can be connected to the wired modem easily so that the user can access the internet without any wireless exposure (providing the wireless function on the laptop is also turned off). On the other hand, a tablet can't be easily wired to the modem, leaving the internet user exposed to wireless radiation from both the tablet and the router.

## Choosing beds

You might wonder what choosing bedding has to do with wireless radiation.

Many mattresses and bases contain metal spring coils. Because metal is a conductor of radiation, some scientists have expressed concerns about the effects of lying on them for long periods of time daily. These researchers measured electric fields from FM radiofrequency signals above beds with and without metal-coil mattresses. They found that, above metal-coil beds, fields were low close to the mattress surface, but they increased with a small distance above it. (This was not the case for beds without metal-coil mattresses.) They even postulated a connection between these fields and breast cancer. Even though the study measured FM radio, which operates at lower frequencies than wireless radiation, it illustrates the conductive nature of wires.

Although the risks of metal-coil mattresses are far from proven, it's probably a good idea to keep metals away from the bed during sleep as much as possible.

If you're in the market for a new mattress, latex, memory-foam, and futon mattresses don't contain spring coils.

## Chapter 11

# Wireless Travellers

How much time do the members of your family spend travelling each week by car or public transport?

If you're like most families, it's probably a considerable amount.

Did you know that cars, buses, trains, and trams all expose travellers to wireless radiation?

The combination of wireless radiation and metal vehicles of all descriptions is quite a potent one. That's because the metal walls reflect the wireless signal so that it bounces around the vehicle, exposing passengers inside.

## Public transport

Some buses, trains, and trams are mini hotspots, providing free wi-fi (in other words, free radiation) to travellers.

In addition, travellers are exposed to radiation from their own and other people's wireless devices.

## Tips for public transport

There's not a great deal you can do to avoid this exposure; however, the following may be useful.

- Identify if your public transport contains a hotspot. If it does, ask the driver where the router is located, and don't sit close to it.
- Avoid adding to your exposure: don't use your own wireless devices when you travel — or use them in airplane mode.
- If you do use a wireless device, don't place it directly against your body while it's connecting to the internet. For example, don't use a laptop directly on your lap; don't have your mobile phone in a pocket or bra when you talk using a headset.
- Travel in a 'silent' carriage, where travellers are less likely to be talking on mobile phones.

Keep in mind that if you want to reduce your exposure to wireless radiation, the most important place to do that is in the home, because you spend far more time there than you do on public transport.

---

**A THOUGHT**

Spare a thought for others. Avoid using your wireless devices on public transport where you can expose other people to the radiation they emit.

---

## Cars

Modern cars can contain an array of Bluetooth-enabled technologies, including mobile-phone kits, music, reversing cameras, GPS, parking sensors, and tyre-pressure monitors. As you've seen, 'Bluetooth' is a type of wireless radiation, so all of these helpful gadgets expose the occupants of the car.

As well as that, travellers are exposed to the fields from all the wireless devices that passengers are using — including any mobile phones that are turned on, irrespective of whether anyone is making a call.

In many places (Australia, the United Kingdom, and over a dozen American states, for example), it's illegal to drive while holding a mobile phone. But did you know that even talking on a hands-free mobile phone while you drive is dangerous, too? Studies have found that talking on a mobile phone while driving reduces driver performance by reducing the driver's ability to position the car on the road, and impairing visual scanning and decision-making — it can quadruple the risk of a crash.

Texting is even more distracting than talking on a mobile phone.

## Tips for car travel

Here is a range of suggestions for reducing exposure to wireless radiation when your family travels by car.

- Don't locate a baby seat near an external aerial.
- Turn off the car Bluetooth.
- Make sure all mobile phones are turned OFF for the duration of the journey.

- Make sure any tablets are in airplane mode.
- Reduce reliance on wireless devices. Keep an activity box in the car to entertain kids. It could contain:
  - plenty of good reading material
  - paper and pencils
  - activity books
  - a pack of cards
  - easy craft (like crochet or weaving).

The advantage of travelling by car is that you have control of the levels of wireless radiation to which your family is exposed.

NOTES

Chapter 12

# Wireless Neighbourhood

When it comes to the safety of your family, you have total control over how much wireless radiation is generated inside your home. You can buy and use an abundance of wireless devices radiating 24/7 — or avoid them altogether. The choice is entirely yours.

However, that's not the case with wireless radiation generated outside your home. Mobile-phone base stations, wireless-internet towers, wi-fi hotspots, smart meters, and even your neighbour's cordless phone and wi-fi can all have an impact on your home.

The extent to which they affect the people inside your home will depend on how closely they're located, the power of the signal they transmit, and, in the case of base stations, whether you're in the main beam of the signal.

Some people use the presence of wireless radiation from external sources as an excuse not to do anything about their internal exposure: *why bother when I'm exposed all the time, anyway?*

The answer is threefold:

1. It's often the case that most of the radiation in a person's home comes from their own wireless devices. Sleeping next to a cordless phone can expose a person to more radiation than from the base station down the road, for example.
2. Reducing your exposure at home reduces your total exposure. This is important because mobile-phone studies are showing that it's the long-term and *cumulative* exposures that are associated with tumour risks.
3. It's relatively easy to block external sources of radiation from entering your home with appropriate shielding. For example: shielding paint, shielding curtain fabric, and shielding window films are readily available from specialist outlets.

Mobile-phone towers, wireless-internet towers, TV towers, and radio towers all emit wireless radiation, although at different frequencies, and, in general, all that's been said about wireless radiation so far also applies to them.

However, there's one technology that deserves a special mention, because of its sheer closeness to your home.

## Smart meters

Like other 'smart' technologies, these electricity meters emit wireless radiation.

One of the problems with these meters is that they're often located on the outside of a bedroom wall, often right near a sleeper's head. As you've already seen, this is not a good position for any wireless device.

Perhaps this is one of the reasons why so many people, in different parts of the world, have reported developing uncomfortable symptoms after smart meters were installed on their homes. Often these symptoms started around the same time as the meters were installed, even when people were unaware they'd been installed.

Melbourne general practitioner Dr Federica Lamech conducted a study of symptoms experienced by 92 residents of the state of Victoria, Australia. The main symptoms reported were insomnia, headaches, tinnitus, lethargy/fatigue, cognitive disturbances, dysaesthesias, and dizziness.

She found that the people surveyed reported symptoms with a remarkable degree of similarity, even though they lived in different parts of the state and had no knowledge of each other.

What can you do?

---

**SUGGESTIONS**

- Reduce the wireless radiation generated by your own devices to reduce your family's total load.
- Don't sleep on the other side of the wall from a smart meter and preferably not in a room that has a smart meter on the wall. If you do, consider shielding the inside wall from the meter to prevent radiation penetrating the room with good quality shielding paint, curtain fabric, or window film.
- Measure the fields inside your home — with all your wireless devices turned off — to see just how much, if any, is coming from outside. You can purchase or hire a good quality meter for measuring radiofrequency radiation. Make sure it comes with appropriate instructions and support materials.

---

# Conclusion

As you take a look at the world around you, it's easy to think that wireless is the way of the future. It's easy to think that teaching your children to use wireless devices will benefit them in later life.

You may be right — wireless devices may be here to stay.

Or their current patterns of use may be a passing phase.

What will determine the longevity of wireless technology is not the advertising that drives sales, not the design of the devices or their functionality. It's the sustainability of the technology — whether its benefits outweigh its risks.

At this stage, we simply don't know if that is the case.

Like any new technology, it will be some time, most likely decades, before we do know whether wireless radiation is safe. The test will be whether the current generation of toddlers who use mobile phones have higher rates of brain tumours as adults, whether children using wi-fi have fertility problems when they reach reproductive age, or whether children's use of these devices damages their behaviour, brain function, relationships, or education.

By then it will be too late to reverse the effects.

As a parent, you can wait for the definitive answers about the safety of wireless radiation — or you can act now to take precautions.

The choice is entirely yours.

Appendix A

# Warnings about Mobile Phones

Here's a handy guide to just some of the many authorities who have recommended that people reduce their exposure to mobile-phone radiation.

### AMERICAN ACADEMY OF PEDIATRICS

The Academy supported congressman Dennis Kucinich's 'Cell Phone Right to Know Act' aimed at investigating the effects of wireless radiation and labelling mobile phones.

### ASSOCIATION FOR CONSUMER PROTECTION (ROMANIA)

The Association released a series of guidelines for reducing people's exposure to mobile-phone radiation.

### ATHENS MEDICAL ASSOCIATION

The Association published '16 Rules for Safer Use of Wireless Communication', including limiting children's use of mobile phones.

### AUSTRALIAN RADIATION PROTECTION AND NUCLEAR SAFETY AGENCY

Australia's radiation authority recommended that parents encourage their children to limit their exposure to mobile-phone radiation.

### AUSTRIAN MEDICAL ASSOCIATION

The Association advised limiting mobile-phone use in children under the age of 16 and suggested methods of reducing exposure.

### BELGIUM

The Belgian government banned the sale of phones designed for children, banned the advertising of phones to children, and required a warning label to be applied to all phones sold in the country.

### BERKELEY, USA

The City of Berkeley passed 'right to know' legislation that required all mobile phones to be sold with this notice: 'If you carry or use your phone in a pants or shirt pocket or tucked into a bra when the phone is ON and connected to a wireless network, you may exceed the federal guidelines for exposure to RF radiation.'

### CANCER ASSOCIATION OF SOUTH AFRICA

The Association recommended minimal use of mobile phones and provided tips for reducing exposure.

### COUNCIL OF EUROPE

The Council of Europe recommended that member states 'take all reasonable measures to reduce exposure to electromagnetic fields, especially to radio frequencies from mobile phones, and particularly the exposure to children and young people who seem to be most at risk from head tumours'.

### DEPARTMENTS OF HEALTH AND EDUCATION (UNITED KINGDOM)

The UK Department of Health issued a statement that children are likely to be more at risk from mobile-phone radiation than adults. Following this, the Department of Education sent a letter to all schools advising children to limit their mobile-phone use.

### DEPARTMENT OF TELECOMMUNICATIONS (INDIA)

The Department announced that it would require mobile-phone retailers to provide customers with a book with information about how to use mobile phones safely.

### FRANCE

The French government introduced bans on the sale of mobile phones to children aged less than six and the advertising of mobile phones to children under 12. It also legislated for an education campaign to inform people how to use mobile devices in a 'responsible and rational' way.

### GERMAN ACADEMY OF PEDIATRICS

The Academy discouraged 'unnecessary, frequent, and extended' mobile-phone use and said children rarely need to use mobiles.

### HEALTH CANADA

Health Canada issued tips for reducing exposure to mobile-phone radiation, and encouraged children under the age of 18 to limit their use of mobiles.

### ISRAEL

The Israeli government requires warning labels with all mobile phones sold, advising that carrying a phone next to the body may increase the risk of cancer, especially for children. It also banned the advertising of mobile phones to children.

### ITALIAN SOCIETY OF PREVENTATIVE AND SOCIAL PEDIATRICS

Italian paediatricians called for a ban on the use of mobile phones by children under ten years of age.

### JAPAN FEDERATION OF BAR ASSOCIATIONS

This group, which represents Japan's 40,000 lawyers, recommended the government take action to limit people's exposure to wireless radiation.

### MINISTRY OF HEALTH (TURKEY)

The Ministry recommended that children and pregnant women reduce exposure to mobile-phone radiation.

### NATIONAL RADIOLOGICAL PROTECTION BOARD (UNITED KINGDOM)

Britain's radiation authority (now called Public Health England) recognised that children could be more vulnerable to mobile-phone radiation.

### PHYSICIANS FOR THE ENVIRONMENT (SWITZERLAND)

These Swiss doctors recommended precautions for mobile phone use.

### PITTSBURGH CANCER INSTITUTE (USA)

The Institute issued advice about how to reduce exposure to mobile-phone radiation.

### RUSSIAN NATIONAL COMMITTEE ON NON-IONIZING RADIATION PROTECTION

Russia's radiation authority warned that mobile-phone use could be as harmful to children as tobacco or alcohol.

### SAN FRANCISCO, USA

The City of San Francisco passed legislation to require all mobile phones to be sold with labels warning of the risks of wireless radiation. However, this ordinance was eventually defeated in court by the mobile-phone industry.

### SENIOR PUBLIC HEALTH SPOKESPERSON

Dr Gro Harlem Brundtland, former physician, prime minister of Norway, international leader in public health, and director-general of the World Health Organization, encouraged children not to use mobile phones.

### TORONTO CITY COUNCIL (CANADA)

Toronto City Council produced a fact sheet with tips about how to reduce exposure to mobile-phone radiation.

### WALES

The Welsh government produced leaflets for teens, recommending a precautionary approach to the use of mobile phones.

Appendix B

# Warnings about Cordless Phones

All of the warnings that apply to mobile phones are relevant to cordless phones. In addition, here are some warnings specifically for cordless-phone use.

### AUSTRALIAN RADIATION PROTECTION AND NUCLEAR SAFETY AGENCY

Australia's radiation authority recommended that parents encourage their children to limit their exposure to cordless-phone radiation.

### AUSTRIAN MEDICAL ASSOCIATION

The Association recommended 'the use of "classical" cord phones' rather than cordless phones.

### BENEVENTO RESOLUTION

Scientists at a 2006 conference in Benevento, Italy, endorsed a statement encouraging governments to adopt precautionary strategies, including the following actions:

'6.2. Inform the population of the potential risks of cell phone and cordless phone use. Advise consumers to limit wireless calls and use a land line for long conversations.

6.3. Limit cell phone and cordless phone use by young children and teenagers to the lowest possible level and urgently ban telecom companies from marketing to them.

6.4. Require manufacturers to supply hands-free kits (via speaker phones or ear phones), with each cell phone and cordless phone.'

### MINISTRY OF HEALTH (ISRAEL)

The Ministry advised people to use a landline rather than a cordless phone. If a person had to use a cordless phone, the Ministry advised using speaker function, keeping the base away from bedrooms and other high-use rooms, and returning the handset to the cradle when the phone is not being used.

### SALZBURG DEPARTMENT OF HEALTH (AUSTRIA)

The Department advised against using DECT (cordless) technologies in schools or kindergartens.

Appendix C

# Warnings about Wi-fi

Again, all of the warnings that apply to mobile phones and cordless phones are relevant to wi-fi. In addition, here are some warnings specifically for wi-fi use.

### ASHLAND, USA

The Ashland school district introduced 'best practice' guidelines for reducing students' exposure to wireless radiation, which are to be displayed in every classroom.

### AMERICAN ACADEMY OF ENVIRONMENTAL MEDICINE

The Academy recommended the use of wired internet and avoidance of exposure to wi-fi and mobile-phone radiation in schools.

### ATHENS MEDICAL ASSOCIATION

The Association recommended limiting wi-fi connectivity and disabling wi-fi on the router at night.

### AUSTRIAN MEDICAL ASSOCIATION

The Association recommended 'Disconnecting (unplugging) the power supply of all WLAN [wireless] access points or WLAN routers.'

### BC CONFEDERATION OF PARENT ADVISORY COUNCILS (CANADA)

The Confederation voted in favour of banning wi-fi in schools in British Columbia.

### BIOINITIATIVE REPORT

This report by a group of independent scientists recommended the use of wired rather than wireless technologies, especially in schools and libraries.

### BORGOFRANCO D'IVREA, ITALY

The municipality of Borgofranco d'Ivrea, a region dubbed the Silicon Valley of Italy, turned off wi-fi in primary and secondary schools due to concerns about children's health.

### CYPRUS

The Cyprian government's National Committee on Environment and Children's Health released a video recommending reducing children's exposure to radiation from wireless devices.

The Minister of Education issued a decree banning wi-fi from preschools, advising that no further wi-fi equipment was to be installed in primary schools, requiring primary schools using wi-fi to switch off the equipment when not in use, and requiring parental consent for wireless activities in schools.

#### FRANCE

No wireless devices can be used in childcare centres for children under three years of age, and wi-fi must be turned off in schools when not being used for study.

#### GERMANY

The German government recommended keeping wireless exposure as low as possible and using wired rather than wireless connections.

#### GHENT, BELGIUM

The municipality of Ghent banned wi-fi from childcare centres used by preschoolers below three years of age.

#### ISRAEL

The Israeli Department of Education advised that wired computer networks should be used in schools.

#### NAVARRA, SPAIN

The government of the autonomous region of Navarra advised that wi-fi should not be used in schools.

#### NEW JERSEY EDUCATION ASSOCIATION (USA)

The Association published guidelines expressing concerns about wireless devices in schools and recommending wired connections for all electronic equipment.

### ONTARIO ENGLISH CATHOLIC TEACHERS' ASSOCIATION (CANADA)

The Association, representing 45,000 teachers, recognised that wi-fi could be a possible workplace hazard, and recommended wiring new buildings for internet connection and introducing methods to reduce exposure to wi-fi radiation.

### PHYSICIANS FOR THE ENVIRONMENT (SWITZERLAND)

In Switzerland, Physicians for the Environment recommend precautions for wi-fi use.

### REYKJAVIK APPEAL ON WIRELESS TECHNOLOGY IN SCHOOLS

Scientists and medical practitioners from 24 countries endorsed this Appeal, calling for wired rather than wireless equipment in schools.

### RUSSIAN NATIONAL COMMITTEE ON NON-IONIZING RADIATION PROTECTION

Russia's radiation authority said that wi-fi radiation is a burden to the child's brain and advised using wired rather than wireless computer networks in schools and other educational institutions.

### SALZBURG DEPARTMENT OF HEALTH (AUSTRIA)

The Department advised that wi-fi signals appeared to be biologically active and advised against using wi-fi or cordless technology in schools or preschools.

### SOUTH TYROL, ITALY

The government of the autonomous region of South Tyrol voted for the replacement of wireless networks with those that emit less radiation.

### SUFFOLK COUNTY, USA

Suffolk County passed a law requiring that notices advise people of the presence of devices that emit wireless radiation.

### TAIWAN

The Taiwanese government legislated to ban parents from allowing children under the age of two from playing with electronic devices.

### VOICE, THE BRITISH TRADE UNION FOR EDUCATION PROFESSIONALS

The union, representing 20,000 lecturers, teachers, and childcare workers, has several times expressed concerns about the use of wi-fi in schools.

# Further Reading

Each claim in this book has been carefully researched and referenced. The comprehensive endnotes can be found online at http://www.emraustralia.com.au/.

If you want to learn more about the fundamentals of living safely in a world of electromagnetic pollution, I explore the research in depth — as well as offering easy-to-understand advice — in my book *The Force*.

NOTES

## NOTES

NOTES